"I wanted to know the name of every stone and flower and insect and bird and beast...."

— George Washington Carver

Benjamin Banneker, American Scientist – Signs of Learning®

— For Young Scientists Everywhere —

Registered Title: Murray David Harwich, III

Text ©2020 Mary Belle Harwich

Illustrations ©2020 Rosalee Anderson

Printed in the United States

Published Frankfort, KY

Book Designs by Marjorie Snelson Design

ISBN 978-0-9888973-1-1

Library of Congress Control Number: 2020917261

To order printed books: www.amazon.com

Benjamin Banneker

American Scientist

1731–1806

Written by

Mary Belle Harwich

Lettered and Illustrated by

Rosalee Anderson

For _____

_____

From _____

_____

Most of all, Benjamin Banneker was curious.
He liked to ask himself questions:
   Where does the wind come from?
   Why does the moon change shape?
   What makes a flower open?
   How does a hummingbird fly backward?

One by one he looked for answers,
   from a tiny insect on a blade of grass,
   to a bird building a nest with winter twigs,
   to his first sight of an evening star —
   and most of all
   answers from his beloved grandmother
   who taught him to read from the family Bible.

He did all of this because, for him,
   life was to learn and to know.

Benjamin took good care of his farm.

Horses, pigs, cows, chickens, and ducks
came running to him when he called.

The corn and wheat grew tall and strong
in his well-tended fields.

Vegetables, apple trees, and honey hives gave
Benjamin and his family healthy and
abundant food.

When Benjamin looked up at night,
   he thought the pattern of stars against the
   darkness was like a giant clock that slowly
   wheeled across the sky toward morning.

When he needed to keep time for himself,
   he took apart a pocket watch, studied its gears,
   then carved a wooden wall clock with a bell
   chime so he could hear it ring, to mark the hours.

His wooden clock ticked steadily along
   for 40 years.

Benjamin's neighbor and friend, George Ellicott,
   shared his many books with Benjamin.

Together they read, and studied and learned
   how to measure land and how to draw maps.

They studied the pattern of sunlight and shadow,
   the flight of wild geese and of bumblebees,
   where butterflies hide when it rains,
   and the seven colors of the rainbow.

Benjamin predicted the exact day and time
    of a solar eclipse,
        knowing the time when the moon would move
        in front of the sun
    to cast a shadow on Earth, making the day
    go dark.

Then as the Moon moved away from the Sun,
    the Earth would once again be flooded with
    light.

His prediction proved true and brought his
    knowledge of astronomy to the attention of
    others.

At the request of President George Washington, Major Andrew Ellicott, cousin of Benjamin's friend George Ellicott, came to Washington to map the land and mark the boundaries of the nation's new capital, Washington, District of Columbia.

Major Ellicott invited Benjamin to go with him as his scientific assistant. Benjamin accepted.

Benjamin's success, not only as a surveyor, but as a kind and helpful man, brought him many new acquaintances in Washington, DC.

When he returned home he wrote a letter to the then Secretary of State, Thomas Jefferson, praising his document, the Declaration of Independence.

This began a lively exchange of letters between Benjamin and Thomas Jefferson, the future president, on justice and the equality of man.

Benjamin wrote and published six almanacs,
   books that told about the weather,
   how to manage land, to keep good records,
   to be mannerly, well organized,
   and useful to yourself and others.

From the almanacs farmers learned
   when to plant their fields
   and when to harvest their crops.

They learned about the fall of rain, and
   how deep the winter snows would be.

They learned how to care for their animals,
   how to grow flowers, and
   when to gather honey from the bees.

And more than anything else,
   the farmers learned how to make their farms
   long-lasting and beautiful.

Benjamin Banneker's
PENNSYLVANIA, DELAWARE,
MARYLAND and VIRGINIA
Almanack
AND
EPHEMERIS,
FOR THE YEAR OF OUR LORD,
1792;

And too, the fishermen learned when
  the ocean tides would rise and fall,
  and how the great waves
  are driven by the wind.

They learned where the silvery fish hide in
  quick-running water and when the
  hard-shelled lobsters would march across
  the ocean floor.

And for the children, too,
  like tiny drops of rain
  he sprinkled funny stories, puzzles, riddles,
  and games into his almanacs.

He wove the weather facts with laughter
  so that children everywhere would
  always love a rainy day.

It could be that when Benjamin finished his
    writing for the day,
    he rested quietly under his favorite tree.

He imagined that his work rested there, too,
    among the strong branches, in the falling
    leaves.

And perhaps he thought that far off in the
    distance, from the house he had built long
    ago he could hear the evening chime of
    his wooden clock.

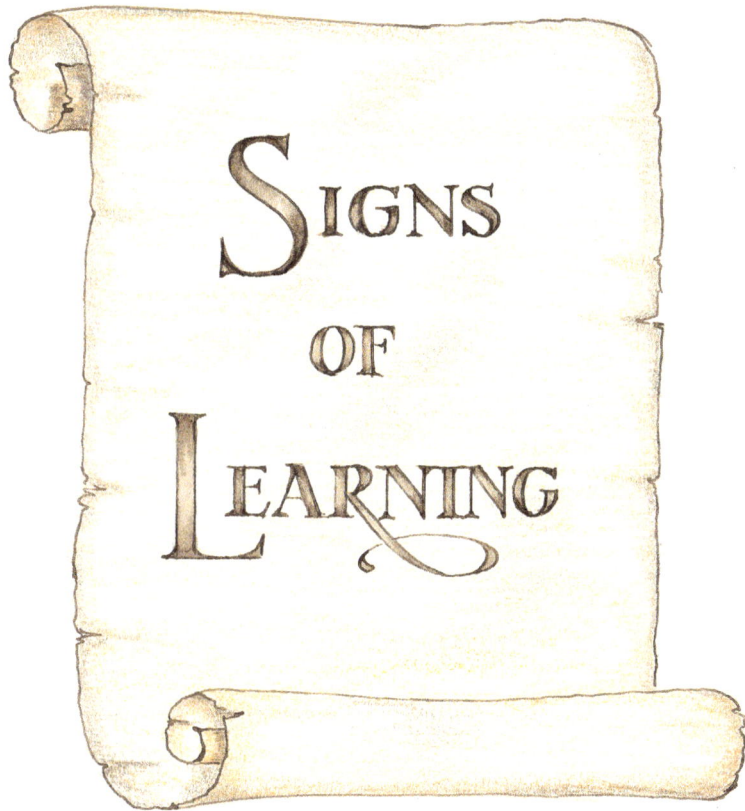

Signs of Learning

Aa

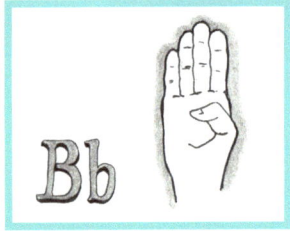

Bb

Cc

Dd

Zz

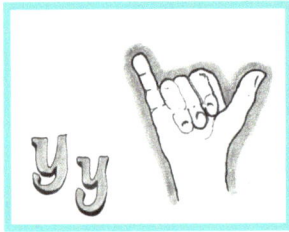

Yy

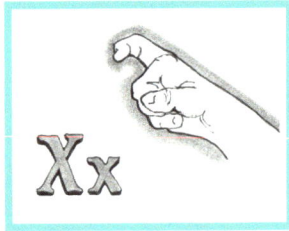

Xx

Ww

Vv

Uu

Tt

Ss

Rr

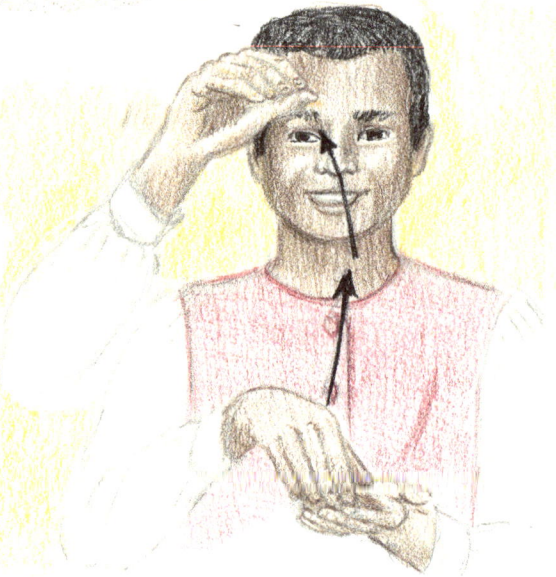

Student

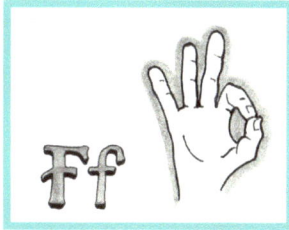

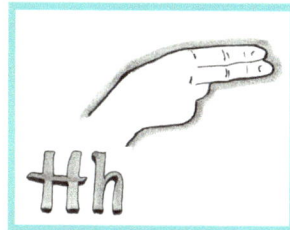

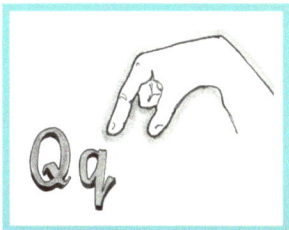

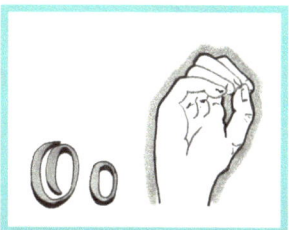

Ee	Ff	Gg	Hh

Ii

Jj

Kk

Ll

Farmer

Mm

Qq	Pp	Oo	Nn

bee

B E E

apple

A P P L E

hen

H E N

chicks

C H I C K S

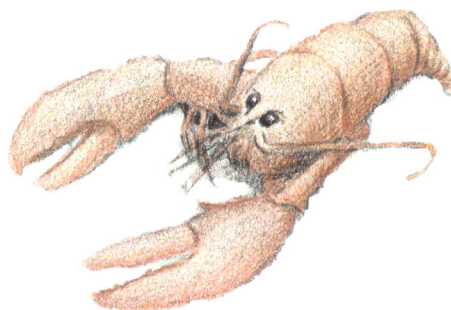

lobster

L O B S T E R

tripod

candle

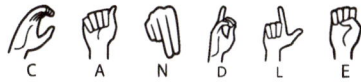

quill

bird

Astronomer

① *star*

② *person*

2x

1 *one*

2 *two*

3 *three*

Surveyor

2x

"Measure"

4
four

5
five

6
six

Author

seven
7

eight
8

nine
9

Nature

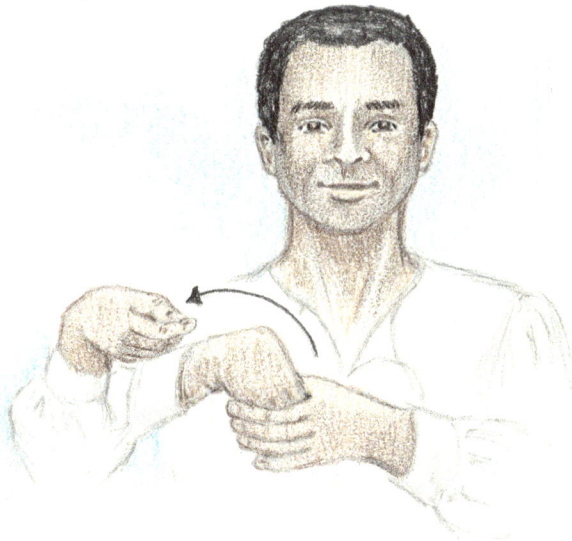

Benjamin Banneker's
PENNSYLVANIA, DELAWARE,
MARYLAND, and VIRGINIA
Almanack
AND
EPHEMERIS,
FOR THE YEAR OF OUR LORD,
1792;

Well Done!

W E L L   D O N E